Quadcopter, Motor and Propeller Calculation

The quadcopters and multicopters used today are driven by brushless motors as standard. In most cases these drive the propellers directly. A gear is used in exceptional cases. This kind of drive is essential to the success of the quadcopter in many technical areas and those of everyday life. It can be implemented easily and without complicated mechanics. In addition, the brushless motors including controllers have efficiencies of over 70% and higher, which also makes the whole drives very efficient. It is therefore to be expected that this type of drive will determine the development of quadcopters for a long time.

In this booklet, the basics for calculating the motor and propeller combinations are worked out. In all cases, these are supplemented with practical calculation examples.

Impressum:

Bibliografische Information der Deutschen Nationalbibliothek:
Die Deutsche Nationalbibliothek verzeichnet diese Publikation in der
Deutschen Nationalbibliografie; detaillierte bibliografische Daten sind im
Internet über http://dnb.dnb.de abrufbar.

© 2021 Roland Büchi

Herstellung und Verlag: BoD – Books on Demand, Norderstedt

ISBN: 978-3-7526-0272-2

Contents

1 Basics of dimensioning

In many cases, only rules of thumb are used for dimensioning. For those who want to do this, the most important laws are summarized right at the beginning:

Roughly speaking, these have their validity even when the theory is examined in greater depth later. With the brushless motors used today, the idle speed is proportional to the applied voltage. The data sheets often contain the parameter kV, which indicates the number of revolutions per minute and V when idling. If you apply a voltage of 10V to a motor with a kV of 1'000 rpm/V, the result is an idle speed of around 10'000 rpm. This is a common value for motors that drive propellers directly and without a gearbox. The mnemonics for quadcopters are the following:

For motors with kV = 1000rpm/V, the following applies:
With a 30g heavy motor you choose an 8"x 4" propeller in direct drive, that results in 4 x 3N = 12N maximum thrust at 10V.
With a 60g heavy motor you choose a 10"x 4" propeller with direct drive, this results in 4 x 6N = 24N maximum thrust at 10V.
With a 100g heavy motor you choose a 12"x 4" propeller with direct drive, that results in 4 x 10 = 40N maximum thrust at 10V.

1.1 Thrust- weight ratio

Typical quadcopter designs have a thrust-to-weight ratio of around 3:1 or 4:1. This means that the maximum achievable total thrust with four motors and propellers is three to four times higher than its own weight.
f a quadcopter in the class of 30g motors and 8" propellers has a total thrust of 1200g or, in physical terms, 12N, then its own weight should not be much more than 400g. At first glance, this ratio appears very high. You'd have to assume that about half of the thrust

should be enough. Tests show, however, that the quadcopter barely hovers when there is no wind.

This is because, on the one hand, you need a so-called controller surplus of around a factor of 2 or 3 in order to have enough power for regulating or balancing gusts of wind, turbulence or other things. With a thrust-to-weight ratio of 2:1, the quadcopter could just barely be floating. If the factor is increased to 3:1 or 4:1, either a little payload can be carried, or speed flight with loops and the like is possible.

This means that about 2/3 to 3/4 is pure reserve, so to speak. In most quadcopters, the motors are openly accessible from all sides and are also located in the propeller air flow. Among other things, this has the very nice advantage that they usually only get lukewarm when hovering and in speed flight. On the one hand, this increases the efficiency, since the motor mainly generates thrust but little heat. On the other hand, it contributes to the long service life.

Of course, the cameras and their mounts also cause additional weight. The smallest cameras weigh only a few grams and can be easily carried by most quadcopters. If everything is to be designed to be automatically rotatable and pivotable and the lens is also of a certain size or offers further possibilities, for example that of a zoom, then the total weight will inevitably increase. Professional systems can soon weigh a kilogram or more. The quadcopter must therefore be designed for this weight in any case.

For a multicopter with a maximum of 1200g (approx. 12N) thrust, the take-off weight including camera should therefore be a maximum of 400g. This could be divided in such a way that the multicopter including battery weighs around 300g and the camera with bracket weighs no more than 100g. A quadcopter with 8" propellers and 30g heavy motors shows this data.

For a multicopter with a maximum of 2400g (approx. 24N) thrust, the take-off weight including camera should not exceed 800g. This could be divided up in such a way that the multicopter including battery weighs around 600g and the camera with bracket weighs no more than 200g. A quadcopter with 10" propellers and 60g heavy motors shows this data.

For a multicopter with a maximum of 4800g (approx. 48N) thrust, the take-off weight including camera should not exceed 1600g. This could be divided up in such a way that the multicopter including battery weighs around 1200 g and the camera with bracket weighs no more than 400 g. An octocopter with 10" propellers and 60g heavy motors shows this data.

For a multicopter with a maximum of 8000g (approx. 80N) thrust, the take-off weight including camera should not exceed 2600g. This could be divided up so that the multicopter including battery weighs around 2000g and the camera with bracket weighs a maximum of 600g. An octocopter with 12" propellers and 100g heavy motors shows this data.

The thrust-to-weight ratio decreases with the payload from 4:1 to around 3:1. There is then still enough regulator reserve to take photos while hovering. If you went below the 3:1 mark, i.e. if you had a little more take-off weight at a certain maximum thrust, then the quad-, hexa- or octocopter could still take off. In the event of small disturbances such as a little wind, however, it may no longer be able to keep itself stable in the balance, since the control reserve required for this is missing. Although in practice one or the other system flies without this reserve, the thrust / weight ratio of 3:1 represents the lower limit of what can be sensibly implemented from the author's point of view. It is always better to use a ratio of 4:1 for flight operations.

1.2 *The basics of propeller dimensioning*

Some of the basic terms used later should be explained first.

Propeller diameter D: the diameter of the propeller, measured from blade tip to blade tip

Propeller pitch H: Since the propeller profiles are twisted, the manufacturers often define the pitch differently. As a rule of thumb, take the propeller pitch at 70% to 75% of D/2 (radius). Then you think to yourself that this is the angle to cut butter for one complete turn. The distance covered by the propeller is the propeller pitch H.

Propeller area F: The area occupied by the rotating propeller:
$F = \pi/4 \cdot D^2$

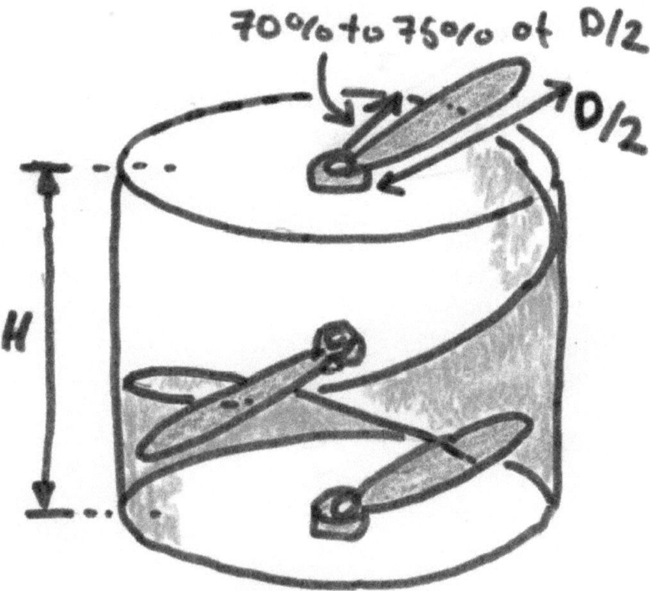

Figure 1: Illustration of the Parameters D und H

In relation to an aircraft, the converted power is equal to the thrust multiplied by the airspeed, according to the formula $P = F \cdot v$. Fast aerobatic models must therefore still be able to develop thrust for figure flight at a flight speed of perhaps 150 km/h.

On the other side are the slow and park flyers. These fly very slowly in comparison. During indoor flights you often see the 'torquing'. Here the model stands vertically in the air. Then there is no speed of the model, only the thrust of the propeller. In this case, the thrust is also called 'static thrust'. Strictly speaking, the performance converted in relation to the model is then also zero. (Static thrust * 0 = 0).

Seen in this way, the quadrocopter belong more to the category of slow and park flyers. Hovering is also a frequent flight situation for them. It is clear that the propellers must also be designed very differently for the different applications.

Compared to slow-flyer propellers, propellers for fast aerobatic models usually have a smaller diameter D with a larger pitch H than slow-flyer propellers. The jet propulsion theory can be used to explain this. It is assumed there that the propeller sucks in the air evenly over the entire propeller surface F. In order for the aerobatic pilot to still generate thrust at the assumed 150 km / h, the air jet must logically be faster than the model itself. This requires a large propeller pitch H. A slowflyer or a quadrocopter does not need a high jet speed, as it is often operated in hover, the pitch H is small. The propeller diameter D can be made slightly larger for the same power. The following rules of thumb apply:

Fast aerobatic models:
ratio of pitch to diameter H / D = 0.7… 1.0
Slowflyer or quadrocopter:
ratio of pitch to diameter H / D = 0.3… 0.7

The above discussed propellers 8" x 4" or 10" x 4" with H / D = 0.5 or 0.4 clearly belong to the 'Slowflyer' or 'Quadrocopter' category. They are designed more for 'hovering' and not 'speed'.

1.3 Specific thrust

For the thrust and the required power, formulas can be derived using the jet propulsion theory. Only the results are presented here. More detailed calculations can be found in the source 'Der Standschub von Propellern und Rotoren'.

$$S = \sqrt[3]{2 \cdot \rho \cdot F \cdot (\xi \cdot \eta \cdot P)^2} \qquad \text{Thrust S (1)}$$

$$P = \frac{1}{\xi \cdot \eta} \cdot \sqrt{\frac{S^3}{2 \cdot \rho \cdot F}} \qquad \text{Power P (2)}$$

ξ: **Propeller efficiency**, 0.5 is the value for an average propeller
η: **Motor efficiency**, it is assumed 0.7 or 0.8
ρ: **Density of the air**, 1.24 kg/m^3
F: Propeller area, as defined above F = π / 4 $\cdot$ D^2

In practice, a common approximate value for quadcopters is 100W required electrical power per kg of hovering mass. After that, every watt of power can make floating 10g of mass. The so-called specific thrust is calculated at 10g/W or technically expressed with the unit 'Newton': S/P = 0.1N/W. This value can now be calculated a little more precisely with the above formulas.

1st example, calculation of specific thrust: The above formulas can be used to calculate how large the specific thrust actually is. Systems with 8", 10" or 12" propellers are calculated. An 800g quadrocopter is taken as an example. An 8" propeller would be too small, remember that it generates a maximum of about 4 x 3N thrust. The weight must be expressed in Newtons, 800g corresponds to about 8N. Since a total of four propellers are involved, the total area is π / 4 $\cdot$ D^2. A 8" propeller has a diameter of 8 $\cdot$ 0.0254m = 0.2032m. At a 10" propeller the value is 0.254m and at a 12" propeller it is 0.3048m.

The calculation with the formula for the power yields interesting things:

8" propeller: P = 114W; S/P = 8N/114W = 0.070 N/W or 7.0 g/W (g are grams)
10" propeller: P = 91.2W; S/P = 8N/91.2W = 0.088 N/W or 8.8 g/W (g are grams)
12" propeller: P = 76W; S/P = 8N/76W = 0.105 N/W or 10.6 g/W (g are grams)

The value of 10g/W assumed above is therefore not far from these values. However, the quadrocopter gets better energetically the bigger the propellers are. If four 8" propellers still need 114W to lift the 800g, the 12" propellers will only need 76W. In practice, however, 12" propellers are also driven by larger motors. So they are heavier. The greater specific thrust is then noticeable by the fact that a heavier quadrocopter with larger propellers requires a battery of the same size as for a smaller one.
Of course, this knowledge applies to the entire flight (model) construction. It is also responsible for the fact that a helicopter has very large rotor blades. At this point one should pause for a moment: "Are quadrocopters that use smaller 8" to 12" propellers, in comparison to model helicopters, a faulty design with regard to the specific thrust?", one might ask. In principle, this could be answered in the affirmative. Since the available Brushless outrunner motors are designed for direct drives of such propellers, the mechanics are simple and there is in most cases no need for a gearbox. This keeps the efficiency high and with normal sized batteries you can still achieve flight times of 10 to 30 minutes, which are comparable with helicopters.

1.4   Propeller with gear

In order to realize a quadrocopter with a larger specific thrust, the complete layout for the use of a 20" propeller of a coaxial helicopter

is shown here. The diameter D is 0.508m. As in the example, it can be used to calculate the hovering power and the specific thrust for an 800g quadcopter.
20" propeller: P = 45.6W; S/P = 8N/45.6W = 0.175 N/W or 17.5 g/W (g are grams)

Compared to the 8" propeller, the required power is 2.5 times smaller! To drive this propeller, one would now like to fall back on the completely normal BL external runners. To do this, however, a gearbox has to be designed.

A typical gear design would be calculated using the desired maximum power at a certain maximum speed. You could then compare this with the maximum speed of a standard combination of 60g motor and 10" propeller and determine the gear ratio accordingly. This method is only partially meaningful here because something else, much more important, has to be taken into account: the mass moment of inertia of the propeller.

Since the control of the pitch and roll angle in quadcopters is realized via the speed, it is essential to ensure that the motor sees the same (preferably smaller, certainly not larger) moment of inertia with the gearbox, as with the standard configuration 60g motor and 10" propeller. This is the only way that the motor / propeller combination is fast enough.
The moment of inertia of a propeller is calculated in a somewhat simplified way like that of a rod with:

$$J = \frac{1}{12} \cdot m \cdot D^2 \quad (3)$$

An example of a standard 10" (D = 0.254m) propeller has a mass of 17g, the weight of the 20" (D = 0.508m) propeller with the blades of a coaxial helicopter is 40g. The two mass moments of inertia are calculated as $93 \cdot 10^{-6}$ kgm^2 und $874 \cdot 10^{-6}$ kgm^2. The moment of inertia of the 20" propeller is around 9.4 times bigger. Gearboxes do not transmit the mass moment of inertia linearly, but quadratically.

In order that the motor sees the same moment of inertia with the 20" propeller, the square of the reduction must be 9.4. The result is a gear ratio of at least 3.07 : 1 (3.07 is the square root of 9.4).

This quadrocopter is a little heavier because of the gearbox, but is hovering longer with the same battery than one with 8" or 10" propellers.

2   Calculation of the propeller speed

The rotation speed is missing in the formulas of the last chapter, so the 'revolutions per minute', with the abbreviation n.

Special coefficients are used to include n in the calculation. The most important ones are the thrust coefficient C_T and the power coefficient C_P. These values are also linked to the propeller efficiency ξ introduced above. The formula comes from the source 'Der Standschub von Propellern und Rotoren'.

$$\xi = \sqrt{\frac{2}{\pi}} \cdot \sqrt{\frac{C_T^3}{C_P^2}} \qquad (4)$$

In order to find propeller efficiencies and coefficients, series of measurements and other theoretical considerations can be carried out; they are carried out in various sources. For the further calculations with a typical slow flyer or quadrocopter propeller in sizes from 8" to 12" and the ratio of pitch to diameter H/D from 0.4 to 0.5, the following values should about apply:

$\xi = 0.5$; $C_T = 0.085$ und $C_P = 0.04$

However, the system is overdetermined according to the above formula; the propeller efficiency ξ is normally calculated from C_T and C_P. For a specific real propeller, its exact data should be used.

Using these values, the static thrust and the power can now also be calculated with the speed. The formulas are from the source 'Der Standschub von Propellern und Rotoren'.

$$S = C_T \cdot \rho \cdot \left(\frac{n}{60}\right)^2 \cdot D^4 \qquad (5)$$

$$P = C_P \cdot \rho \cdot \left(\frac{n}{60}\right)^3 \cdot D^5 \qquad (6)$$

2nd example, calculation of thrust and power: For a quadcopter with a 10" x 4" propeller and n = 4000 rpm, thrust and power should be calculated. There is D = 0.254m, ρ = 1.24 kg/m³ ,C_T = 0.085 and C_P = 0.04, so the total thrust is 4 · 1.95N = 7.80N. At this speed, about 800g weight can be kept hovering. The power P is 4 · 15.54W = 62.15W according to the above formula. But this is about the propeller power. If you want to calculate the total electrical power of the motors, this value has to be divided by the motor efficiency η. If this is assumed to be 0.7, as in the first example, the electrical power is 62.15W / 0.7 = 88.78W. This value is slightly lower than there (see 91.2W), but the thrust is also slightly lower due to the use of a 'round' speed, and the values C_T and C_P are also slightly rounded.
The current that flows in this case is calculated for a three-cell LiPo battery at 88.78W / 11.1V = 8A.
If one also assumes that the quadrocopter has a 4000mAh battery, then a flight time of 4Ah / 8A = 0.5h, half an hour is the result.

3rd example, calculation of the maximum thrust: The maximum thrust should now be calculated for the same quadrocopter. Assuming that the motor has a kV 1'000 rpm/V and a maximum of 10V is applied, then the maximum speed is 10'000 rpm. If you use these numbers, the result is a maximum thrust of 50N or about 5kg. This value is about twice as high as stated in the clauses above, there is only talk of 4 x 600g = 2400g. In fact, at the applied 10V, the motor no longer brings 10'000 rpm. Because of the high current and, in some systems, also because of the software limitation, it stops at around 6'000 to 7'000 rpm. In order to calculate this more precisely, the motor characteristic curve is also included below. If, however, the value 7'000 rpm is used in the formulas, the maximum thrust is 4 · 5.97N = 23.88N. The electrical power of the motors with the calculated efficiency is 4 · 119W = 476W. The maximum current for a three-cell LiPo battery is 476W / 11.1V = 42.9A.

When examining the formulas for thrust and power, an interesting mnemonic can be found:

The thrust changes with the square of the speed, while the required power changes with the third power of the speed.

In other words: the higher the speed, the smaller the specific thrust, since the power then increases more than the thrust.

4th example, for the 2nd und 3rd example (same motors and propellers, with 4000 rpm respectively 7000 rpm), the specific thrust should be calculated:
For the 2nd example, S/P = 7.8 N / 88.78W = 0.088 N/W or 8.8 g/W (g are grams)
For the 3rd example, S/P = 23.88 N / 476W = 0.050 N/W or 5 g/W (g are grams)
So the mnemonic is confirmed.

The motor mass is a measure for the power. Motors of the 30g weight class should be designed for a maximum power of no more than about 75W, motors weighing 60g for a maximum of about 150W. Therefore for the above examples, 60g heavy motors should be used. However, since mostly only the "floating power" from the 2nd example is required, they are greatly oversized with 92W / 4 = 23W continuous power. On the one hand, the rest is reserve, just think of the thrust-to-weight ratio of 4:1 discussed above. On the other hand, the motors are usually completely in the propeller air flow due to the assembly. In contrast to their use in model airplanes, where they are often operated at maximum power, they usually only get lukewarm here in flight.

3   Brushless motor calculation

In the examples above, only the propeller was used for calculations only. In particular, the maximum motor speed in operation was only estimated. In order to take this into account, the motor characteristic must also be included in the calculation.
The two formulas for calculating static thrust and power are repeated for better understanding.

$$S = C_T \cdot \delta \cdot \left(\frac{n}{60}\right)^2 \cdot D^4 \qquad (5)$$

$$P = C_P \cdot \delta \cdot \left(\frac{n}{60}\right)^3 \cdot D^5 \qquad (6)$$

There are:
S: thrust, C_T: thrust coefficient, δ: density of the air, 1.24 kg/m^3, n: revolutions per minute, rpm, D: propeller diameter, P: mechanical power on the shaft, C_P: power coefficient.

The mechanical power is calculated by P = n · M · π/30 (or P = M · ω, with ω = 2 · π · f = 2 · π · n/60). This means that the propeller torque can also be determined from the formula for the power.

$$M = C_P \cdot \delta \cdot \frac{1}{2 \cdot \pi} \cdot \left(\frac{n}{60}\right)^2 \cdot D^5 \qquad (7)$$

The rotational speed is therefore included in the formula as a square. A comparison with Figure 3 shows that the speed- torque characteristic of the propeller is thus parabolic. This is the case with quadratic relationships. The diameter D is also interesting. It is also contained in the formulas for torque and power with the 5th power and in the formulas for static thrust with the 4th power. This means that power, torque and thrust increase massively at a larger diameter.

3.1 Formulas of the brushless motor

The brushless motor has a linear speed- torque characteristic. In principle, it can be described with the electrical equivalent circuit diagram below as shown in Figure 2.

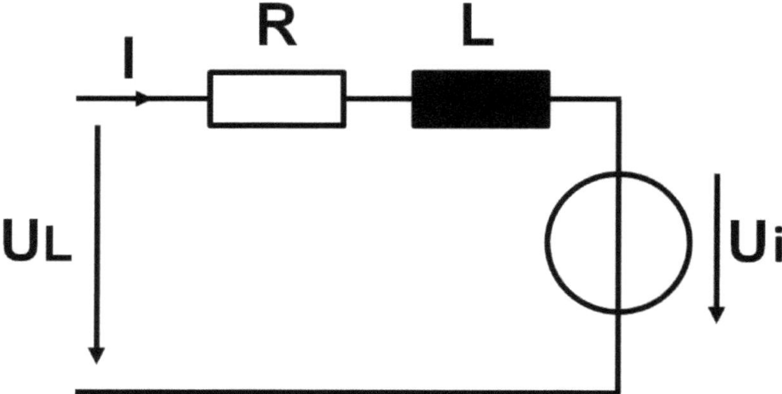

Figure 2: Electrical equivalent circuit of the brushless motor

U_L refers to the voltage that is applied between the conductors. This is provided by the brushless controller. At full throttle, U_L is equal to the battery voltage U_{BAT}, at other positions the brushless controller with PWM provides a correspondingly lower voltage. Ui is the induced counter voltage. The following applies to the speed:

$$n = (U_L - R \cdot I) \cdot kV \quad (8)$$

The motor constant kV is the most important parameter of the motor. kV is usually given in revolutions per minute and volt, the two formulas 8 and 9 also take these units into account. When idling, the motor current is I = I₀. The motor inductance L can be neglected for the considerations made here. As soon as the load generates a torque, a current I flows. The relationship between the torque and the current is calculated by

$$M = (I - I_0) \cdot km, \quad km = \frac{1}{kV} \cdot \frac{30}{\pi} \qquad (9)$$

With this information, the speed- torque characteristic of the motor can be drawn. The higher the torque, the higher is the motor current and the lower the speed. The efficiency assumed in the calculations of earlier chapters can also be calculated for each operating point of the motor; this is the ratio between the mechanical and the electrical power.

$$\mu = \frac{P_{mech}}{P_{el}} = \frac{(U_L - I \cdot R) \cdot (I - I_0)}{U_L \cdot I} \qquad (10)$$

An example of a motor and propeller characteristic is shown in Figure 3. The intersection of the two characteristic curves is the operating point.

5th example, motor and propeller calculations with a constant rotational speed: The brushless controller keeps a 10" propeller at a constant 5'000 rpm. A motor current of 8 A and a static thrust of 5 N (corresponds to approx. 500 g) are measured with a scale (see figure 4). A 12" propeller with the same ratio of pitch to diameter (H / D) is now screwed onto the motor. Here, too, the brushless controller keeps the speed at a constant 5'000 rpm. The new motor current and the new static thrust are to be calculated

A propeller of the same design with a similar H / D also results in similar power and thrust coefficients C_P and C_T. As is known from chapter 3.1, the motor current I is proportional to the torque M. Since in the above formulas for the torque and the static thrust all quantities except the diameter remain the same, the following applies: I_new = I_old · D_new^5/D_old^5 = 8 A · 0.3048^5 / 0.254^5 = 12.4416 A. The same applies to the static thrust: S_new = S_old · D_new^4/D_old^4 = 5 N · 0.3048^4 / 0.254^4 = 10.37 N (about 1 kg).

The example nicely shows the influence of the 5th or 4th power of the diameter. When a 10" propeller is replaced by a 12" propeller, both the static thrust and the motor current are more than doubled. If the speed is not kept constant, it will be slightly smaller because of the larger current and the voltage drop at the internal resistance R, and the new current or the new static thrust will not be quite as high as in the calculated example. The order of magnitude "factor two" will still be about right.

6th example, complete calculation of a quadcopter:
A quadrocopter is supplied with a 3S LiPo battery (= 11V). Both the battery and the brushless controller are assumed to be lossless. The coefficients found for a 10 inch propeller are approximately C_P = 0.035 and C_T = 0.08. The motor has a kV of 924 rpm/V, a winding resistance R of 0.089 Ω and an idle current I_0 of 0.75 A.
A quadrocopter is to be completely calculated from this.
The formula for the motor characteristics is: $n = (U_L - I · R) · kV$. The formula for the propeller curve (load curve) is: $M = C_P · \delta · 1/2\pi · (n/60)^2 · D^5$. D is 0.254 m. There is also $M = (I - I_0) · k_m$, with $k_m = 1/kV · 30/\pi = 1/924 · 30/\pi$ Nm/A = 0.0103 Nm/A.
The required operating point can only be found numerically or graphically due to the non-linear propeller characteristic. Figure 3 shows the motor characteristic and the propeller characteristic as well as their intersection. The operating point for the maximum rotational speed is calculated as

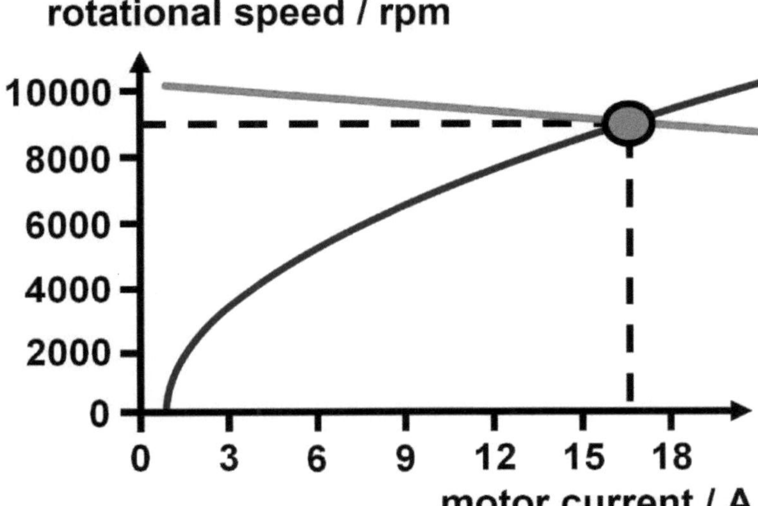

Figure 3: Motor and propeller characteristics

For the point of intersection (operating point) one finds about n = 8850 rpm, I = 17 A. That gives a torque $M = (I - I_0) \cdot km = (17-0.75) \cdot 0.0103$ Nm = 0.167 Nm.

The static thrust can also be calculated using the static thrust formula. $S = C_T \cdot \delta \cdot (n/60)^2 \cdot D^4 = 0.08 \cdot 1.24 \cdot (8850/60)^2 \cdot 0.254^4 = 8.98$ N (or about 900 g). The electrical power of the motor is $P_{EL} = U_{Bat} \cdot I = 11.1$ V $\cdot$ 17 A = 187 W. The mechanical power is $P_{MECH} = n \cdot M \cdot \pi / 30 = 8850 \cdot 0.167 \cdot 3.1415 / 30$ W = 155 W. The motor efficiency is then $\eta = P_{MECH} / P_{EL} = 155$ W / 187 W = 82.9 %.

Above, the rule of thumb for the thrust / weight ratio was 4: 1. In total, 4 propellers deliver a maximum of 8.98 N thrust each, i.e. 35.92 N. This means that this quadrocopter can hover with 35.92 / 4 = 8.98N, which corresponds to a flight weight of around 900 g.
Each drive brings 8.98N / 4 = 2.245N thrust when hovering.

Now formula 5 has to be solved after the speed n. This results in: $n = 60 \times$ sqrt $(S / (C_T \cdot \delta \cdot D^4) = 60 \times$ sqrt $(2.245 / (0.08 \cdot 1.24 \cdot 0.254^4)$ rpm $= 4'424$ rpm. In hover flight, this drive has a mechanical power $P = C_P \cdot \delta \cdot (n/60)^3 \cdot D^5 = 0.035 \cdot 1.24 \cdot (4424/60)^3 \cdot 0.254^5 = 18.39W$. With an assumed motor efficiency of 0.8, the electrical power is 18.39 W / 0.8 = 23W. With four drives this results in 92 W. The current is 92W / 11V = 8.36A.
Assuming that you can mount a LiPo battery with 4000 mAh, the result in hover is a flight time of 4Ah / 8.36A = 0.478 h or 28.7 minutes.

In the present example, it should be noted that the propeller coefficients of the propellers available in practice are rather larger than the assumed data. Thus, the flight duration is calculated rather conservatively.

The calculation from example 6 can today also be carried out by programs on the Internet. These can also take into account the internal resistance of the battery and the brushless controller. The relevant total resistance is ultimately just a pure addition of the single internal resistances. In addition, the data from many propellers, motors, controllers and batteries are already saved. It is therefore possible to calculate online a complete drive unit including the propeller. The author, who has already compared many calculations and measurements and also designed and built many quadrocopters himself, would estimate the deviations from reality to be below 10%.

4   Three or four blade propellers

The source "Der Standschub von Propellern und Rotoren" also provides formulas for using the three- or four-blade propellers.

For a three-blade propeller, the thrust increases about by a factor of 1.4 at the same rotational speed and the power increases about by a factor of 1.6, compared to the two-bladed propeller of the same design.
For a four-blade propeller, the thrust is increasedabout by a factor of 1.8 and the power is increased about by a factor of 2.2, compared to the two-blade propeller of the same design.
There are very few quadrocopters that are powered by three or four blade propellers. This only makes sense if they have to be built compactly and only a small amount of propeller space is available. The reason for this can be found here. The required power always increases at the same rotational speed more (by 1.6 or 2.2) than the generated thrust (by 1.4 or 1.8). With the same diameter, such propellers are always less efficient than the two-bladed propellers.

5   Power and thrust measurement

It is possible with a voltmeter and a normal kitchen scale to take both measurements in the title.
In the previous chapters, the total output was always calculated. Since the consumption of the control electronics of the quadcopter is low compared to the motors, the power can be measured directly on the battery. The voltage measurement is not a problem here. But the currents, which are in the range between 5A and 50A, can no longer be measured with commercially available multimeters, as these are rarely designed for more than 10A. A shunt resistance of 0.01Ω, in series between the battery and control electronics, is the solution here. With a current flow of 20A, $20A \cdot 0.01V = 0.2V$ is measured at its connections. The voltage drop is only minimal

compared to the 10V of the battery and only slightly falsifies the measurement. The shunt should be designed for a few watts, otherwise it will get too hot with the high current. As an example, a 3S Lipo battery then delivers the output 20A · 11.1V = 222W at a current of 20A.

Since the center of the quadrocopter does not generate any thrust, it can easily be placed on a kitchen scale. You have to make sure that it is weighted down well so that nothing can ever slip. You also have to make sure that no propeller blows on the scales, otherwise the results will be falsified. The weight difference between full throttle and standstill is then the maximum thrust. If possible, the system should be kept away a propeller diameter away from the ground, otherwise the measured thrust will be a little too high due to the ground effect.

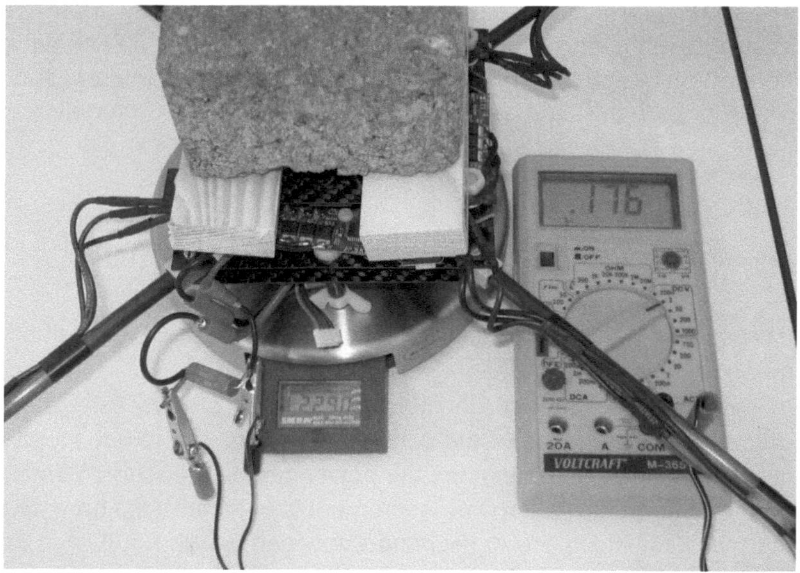

Figure 4: Thrust measurement using a kitchen scale, current measurement with a multimeter and a shunt resistor. The current is here: 0.176 V / 0.01 Ω = 17.6 A

6 Calculation of the propeller coefficients

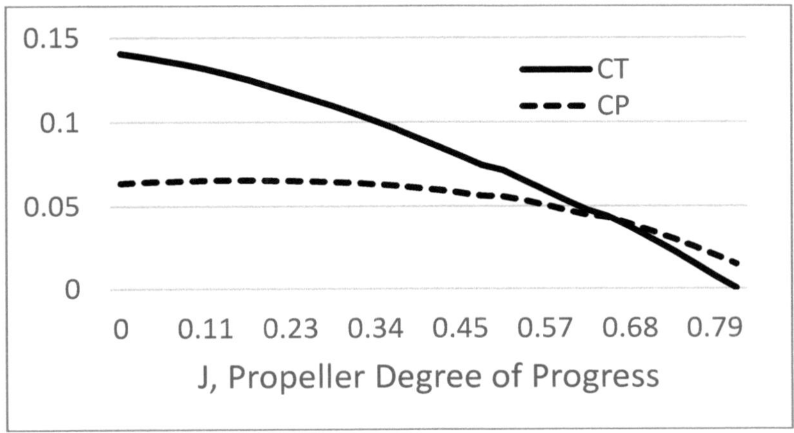

Figure 5: C_T and C_P of the propeller APC SF 10" x 4.7" at 2000 rpm (Source: data download, March 2021)

In many cases, the propeller coefficients C_T and C_P are available from the manufacturers as series of measurements. It should be noted, however, that these are dependent on the speed of the quadcopter or the flight system in general. This dependency is represented with the degree of progress J.

For the calculations made here, only the hovering flight of the quadrocopter is taken into account. So for this consideration   J = 0 and the values of C_T and C_P for the propeller shown in the picture are approximately 0.14 and 0.06. Since the speed also influences the degree of progress, these curves must, strictly speaking, be measured for all speeds. It turns out, however, that the values in hover flight are very similar over a very large speed range. For this reason, they can be assumed to be constant to a good approximation in the practical calculation.

When designing drives, it is important to point out that the manufacturers specify maximum speeds for their propellers. These must not be exceeded.

The formulas (5) and (6) can be solved for C_T and C_P, respectively.

$$C_T = \frac{S}{\delta \cdot \left(\frac{n}{60}\right)^2 \cdot D^4} \qquad (11)$$

$$C_P = \frac{P}{\delta \cdot \left(\frac{n}{60}\right)^3 \cdot D^5} \qquad (12)$$

7th example, calculation of the propeller coefficients from a thrust measurement:

At the measurement of a quadrocopter, a total of 7A flows with a LiPo battery of 11V. The propeller has a diameter of 11 inches, as propeller speed it is measured 3000 rpm and the scale measures a thrust of 10N (approx. 1 kg). The efficiency μ of the motor is 0.7. Calculate the propeller coefficients C_T and C_P and the efficiency of the propeller ξ.

The electrical power Pel is calculated as 11V·7A = 77W, assuming that the electronics of the system are lossless. The mechanical power P of each individual drive is then Pel · μ / 4 = 77W · 0.7 / 4 = 13.475W.

The diameter D is 11 · 0.0254m = 0.2795m.

Thus becomes C_P = P/(δ·(n/60)^3·D^5) = 0.051.

With a thrust of the four drives of 10N each one makes 2.5N. The value C_T = S/(δ·(n/60)^2·D^4) = 0.106. The efficiency of the propeller is calculated using formula (4) as 0.54.

7   Summary

Some 'Lesson's learned'

The same propellers are used for quadrocopters as for slow flyers. The ratio of pitch to diameter H / D is usually between 0.4 and 0.5.

The specific thrust increases with the larger propeller diameter. Quadrocopters with propellers that are smaller than helicopters only achieve attractive flight times because they can be operated with direct drive BL external rotor motors, which are highly efficient.

The faster the propeller turns, the smaller the specific thrust, since the power increases with the third power, but the thrust only increases with the second power of the rotational speed.

Motors weighing 30g should be designed for a maximum output of 75W, 60g motors for a maximum output of 150W. For motor with other weights, the maximum power changes accordingly.

8 Literature

(1) Schenk, Helmut. "Der Standschub von Propellern und Rotoren." (2007).

(2) Wendel, Jan. *Integrierte Navigationssysteme: Sensordatenfusion, GPS und Inertiale Navigation*. Oldenbourg Verlag, 2009.

(3) Grewal, Mohinder S., Lawrence R. Weill, and Angus P. Andrews. *Global positioning systems, inertial navigation, and integration*. John Wiley & Sons, 2007.

(4) Büchi, Roland. *Radio control with 2.4 GHz*. BoD–Books on Demand, 2014., ISBN 978-3732293407

(5) Passern, Ulrich. *Das LiPo-Buch: Grundlagen und Praxistipps-Aktualisierte Neuauflage*. Verlag für Technik und Handwerk, 2016.

(6) Föllinger, Otto, Frank Dörrscheidt, and Manfred Klittich. "Regelungstechnik. Einführung in die Methoden und ihre Anwendung." *Berlin: Elitera Verlag* (1978).

(7) Büchi, Roland. "Brushless-Motoren und–Regler. 1." *Auflage, Baden-Baden: Verlag für Technik und Handwerk neue Medien GmbH* (2013).